by Stuart J. Murphy • illustrated by Jon Buller

HarperCollins*Publishers*

To M.E.M.—who made sure that I was
ready, set, and raring to hop.
—S.J.M.

The illustrations in this book were done with ink and watercolor on Fabriano paper.

For more information about the MathStart series, please write to
HarperCollins Children's Books, 10 East 53rd Street, New York, NY 10022.

Bugs incorporated in the MathStart series design were painted by Jon Buller.

HarperCollins®, ✦®, and MathStart™ are trademarks of HarperCollins Publishers Inc.

Library of Congress Cataloging-in-Publication Data
Murphy, Stuart J., date
    Ready, set, hop! / by Stuart J. Murphy ; illustrated by Jon
Buller.
        p.   cm.   (MathStart. Level 3)
    Summary: Explains equation building as two frogs count
their hops to a rock, a log, and a pond.
    ISBN 0-06-025877-2. — ISBN 0-06-025878-0 (lib. bdg.)
    ISBN 0-06-446702-3 (pbk.)
    1. Addition—Juvenile literature.   2. Subtraction—Juvenile
literature.   [1. Addition.   2. Subtraction.]   I. Buller, Jon,
date,   ill.   II. Title.   III. Series.
QA115.M874   1996   95-4729
513.2'11—dc20   CIP
                                                      AC

Typography by Elynn Cohen
1   2   3   4   5   6   7   8   9   10
❖
First Edition

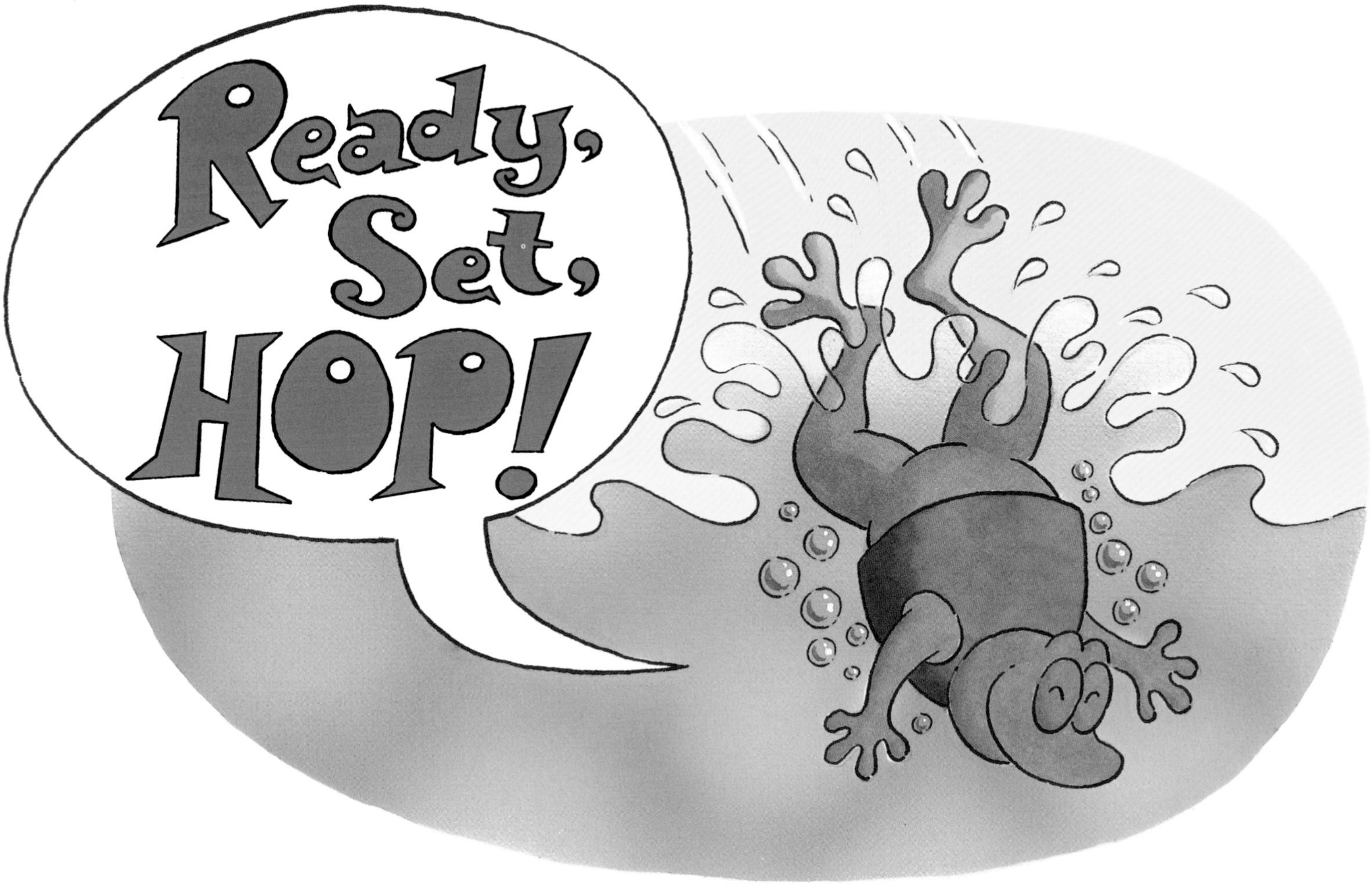
Ready, Set, HOP!

Matty and Moe are two happy frogs.

They like to play on rocks and make
rafts from tiny logs.

One day Matty bragged to Moe, "I'm a very tall frog. I bet I can hop longer hops than you."

"But Matty," croaked Moe, "I'm a very big frog, and I hop long hops too."

"Let's find out," said Matty. "I bet I can get to the big rock in fewer hops than you, because my hops are longer."

7

"But my hops are bigger," said Moe. "I'll
go first and show you."

He stretched his large legs and started
hopping.

It took him 5 hops to get to the big rock.

Moe croaked, "Now see if you can beat
that. READY, SET, HOP!"

It took Matty 2 hops more than it took
Moe to get to the big rock.

11

How many hops did it take Matty to get
to the rock?

It took Moe 5 hops.

It took Matty 2 hops more than it took Moe.

5 plus 2 equals how many?
5 + 2 = ?

"I can't believe it," said Matty. "You took fewer hops, so you're ahead so far. But now let's hop to the hollow log."

Moe wasn't worried. He knew he could hop longer hops.

It took him 7 hops to get to the log.

"Okay, Matty, it's your turn—
READY, SET, HOP!"

Matty hopped 3 hops fewer than Moe to get to the log.

How many hops did it take Matty to get
from the rock to the log?

It took Moe 7 hops.

It took Matty 3 hops less than it took Moe.

7 minus 3 equals how many?

7 − 3 = ?

Matty said, "We were each ahead one time, so we'd better try once more."

"Okay," said Moe. "Watch how few hops I take to get to the pond."

"After you, Moe," said Matty.

With 7 long hops, Moe reached the pond.

"Come on, Matty," yelled Moe. "READY, SET, HOP!"

Matty hopped and hopped. He hopped 2 hops more than Moe—but he should have stopped and hopped 1 less.

How many hops did Matty take to get from the log to the pond? How many should he have taken?

It took Moe 7 hops.

It took Matty 2 hops more than it took Moe—but he should have hopped 1 less.

7 plus 2 equals how many?

minus 1 equals how many?

$7 + 2 = ? - 1 = ?$

Moe croaked, "The contest is over. Let's
add up all our hops and see who won."

Moe counted: "I hopped 5 hops to get to the rock, then 7 more to the log—and 7 more to the pond. That gives me 19 hops."

26

Matty said, "I hopped 7 and 4 and 9. That makes 20 hops in all. It looks like you finished with one hop less than me. So Moe, you're the better hopper."

"I won! But all that hopping made me hot," croaked Moe. "Watch out! I'm hopping in."

Matty said, "If you take one more hop,

19 + 1 = 20

then both of us will win!"

If you would like to have fun with the math concepts presented in *Ready, Set, Hop!*, here are a few suggestions:

- Read the story with the child and describe what is going on in each picture. Point out where the information needed to answer each of the questions appears within the story.

- Identify the clues that suggest operations—"more" suggests addition, "less" suggests subtraction.

- Encourage the child to tell the story using the math vocabulary: "More," "Plus," "Fewer," "Less," etc.

- Reread the book together and ask the child to count the jumps that Matty and Moe take. Try to solve the problems before reaching the math summary pages.

- Gather some crayons, buttons, or keys. Ask, "How many are in the pile?" Add a few more. Ask, "How many are in the pile now?" Take away several and ask, "Now how many are there?"

- Look at things in the real world and work together to create addition and subtraction problems. Examples could include fruit: 3 apples plus 2 oranges equals 5 pieces of fruit, or pets: 3 dogs minus 1 dog equals 2 dogs. Draw pictures of these examples and write the equations under the pictures.